YOUR KNOWLEDGE HAS VALUE

Suleiman Usman

Agricultural Soil Environment

GRIN Verlag

Bibliografische Information der Deutschen Nationalbibliothek:

Die Deutsche Bibliothek verzeichnet diese Publikation in der Deutschen National-
bibliografie; detaillierte bibliografische Daten sind im Internet über http://dnb.d-
nb.de/ abrufbar.

Dieses Werk sowie alle darin enthaltenen einzelnen Beiträge und Abbildungen
sind urheberrechtlich geschützt. Jede Verwertung, die nicht ausdrücklich vom
Urheberrechtsschutz zugelassen ist, bedarf der vorherigen Zustimmung des Verla-
ges. Das gilt insbesondere für Vervielfältigungen, Bearbeitungen, Übersetzungen,
Mikroverfilmungen, Auswertungen durch Datenbanken und für die Einspeicherung
und Verarbeitung in elektronische Systeme. Alle Rechte, auch die des auszugsweisen
Nachdrucks, der fotomechanischen Wiedergabe (einschließlich Mikrokopie) sowie
der Auswertung durch Datenbanken oder ähnliche Einrichtungen, vorbehalten.

Imprint:

Copyright © 2013 GRIN Verlag GmbH
Druck und Bindung: Books on Demand GmbH, Norderstedt Germany
ISBN: 978-3-656-36913-4

This book at GRIN:

http://www.grin.com/en/e-book/208742/agricultural-soil-environment

AGRICULTURAL
SOIL ENVIRONMENT

S. Usman

AGRICULTURAL SOIL ENVIRONMENT

.

SULEIMAN USMAN
Natural Resources Institute,
The University of Greenwich
Medway Campus, Chatham Maritime
Central Avenue, Chatham, Kent ME7 4TB UK

Content

Summary

It is believe that a better understanding of agricultural soil environment is a key to better soil quality and high crop yield. Understanding the soil problems would certainly leads to solutions on how to tackle them. However, this could not only serve as a part of soil management but also as a part of economic development. For instance, when mother is healthy and fertile, there is more expectation from her to provide healthy and quality produce but when there is deficiency in any of the elements (either macros or micros elements), her produce may also affected, hence increasing poverty and hunger. High quality and fertile soil always provide better condition for the growth and development of plants (right from seed germination up to the maturity stage). Example, deficiency in N, P, and K in soil might lead to several problems to growing plant in a particular soil environment. These problems may include poor germination, stunting, and defoliations of plants, late maturity and also susceptible to both pest and disease attack leading to poor yield of the crops. Therefore, understanding soil environment is a key to high yield of crop production in the northwest one of northern Nigeria.

1.0 Introduction

Soil environment is a significant component of earth surface nature and as such provides many benefits to mankind. These benefits include agriculture, housing, vegetation, water and inter-communication between one geographical region to another. In agriculture, one of the most important benefit that soil environment offer is make available of moisture and nutrients contents for the basic requirement of plant growth (Russell and Greacen, 1977; Zachar, 1982; Yost *et al.*, 1987; Wild, 1993; Brady and Weil, 2004; Usman, 2013). However, for many years soils in the globe have been facing lots of problems starting from nutrient depletion, as a result of land degradation factors such as erosion by wind and water, desertification as a result of deforestation or other related factors, poor vegetation cover as a result of human impact on natural resources such as human induced activities of destroying forest. These types of problems have lead to grave damages to regional soil environment, agricultural land cover and food availability as a result of organic matter lost or deterioration of soil structural quality and textural quality as well. Therefore, the objective of this synopsis paper is to provide a short note reference on soil environment as existing information in the context of north-west zone of northern Nigeria.

2.0 Soil environment in north-western Nigeria

Northern Nigeria can be divided in to two important geographical zones namely: the north-east zone and north-west zones. These two zones are dominated predominantly by the Hausa-Fulani people, who depend on farming and cattle rearing. The Federal Ministry of Environment Nigeria (FMEN, 2002) has indicated that the north-west zone included Kebbi, Sokoto and Zamfara States. These States bordered by the republic of Niger and Chad (Aregheore, 2005). The north-west zone covers a wide range of land described as the dryland of Nigeria, which constitute the main source of fodder and grazing land for livestock (Gadzama, 1995; Usman, 2007).

The north-west zone of Nigeria has an undulating plain topography characterised by high sand particles, usually very low in organic matter (Aregheore, 2005). The surface soil of the dryland area in the region forms an undulating plain at a general elevation covering about 450 to 700 m, and more than half of the region is covered by the ferruginous tropic soils, which are highly weathered and markedly particles (Mortmore, 1989). A large proportion of the region is also characterised by sandy-fixed undulating topography. The sandy soil is usually low in organic matter, nitrogen and phosphorous and may degrade rapidly under conditions of intensive rainfall.

Soils in the north-west-zone are well drained, with low water-holding capacity but having a deep water table in most of fadama area (Usman, 2003). The fadama flood plains areas of the region can be described as having flat-floored valleys that are flooded in the wet season only, and recedes during the dry season and leaves a coating of alluvial soil well covered. These seasonally flooded areas are called "fadama" in Hausa (Iloeje, 2001). The fadama soils have stable textural soil condition, and this largely controls the shape and size of pore space in soil, as such governs the infiltration capacity and run-off (Datta, 1986). Two soil textures are important in this circumstance – the fine and coarse-texture. Fine-textured soils have mostly narrow capillary channels; while coarse-textured soils have large and wide non-capillary pores; thus, the finer the texture the greater is the percentage of run-off (Usman, 2013).

2.1 Key soil types and environmental soil conditions

The key soil type in the north-west zone of northern Nigeria can be physically classified according the agricultural site as Aridisols in dryland site and Vertisols in fadama site (USDA, 1975). The dryland Aridisols are characterised as slowly permeable and most of the water is lost by run-off (Fitzpatrick, 1980). They might also have been formed under aridity from wind stored desert sands that accumulated over long period of time when the Sahara desert encourage several kilometres south of its present limits (Adegbola, 1979). The fadama soils in contrast, hold more water within the very-tight textural pores and have low run-off. They might have formed under flood plain nature covered largely with alluvial textural particles.

In addition, the dryland soils in the north-west zone might be also qualified to ferruginous tropic soils (D'Hoore, 1964); these ferruginous tropic soils can be characterised by having high sandy textural particles covering large areas of land, very low water holding capacity and low organic matter, low nitrogen content, low available phosphorus, neutral or moderately acid in pH (Jones and Wild, 1975). Thus, their ability to hold nutrients for plant growth in this circumstance is also very week.

Physically, the vegetation is usually spare and the surface is bare for long period. In essence, this may have contributed a lot to soil degradation by wind erosion, causing soil fertility decline in the zone. However as a matter of fact, this is one of the contributing factors to great soil environmental problem under agricultural condition in the north-west zone of northern Nigeria; and also making greater support to land degradation, which is also a worldwide phenomenon.

Physically, in north-west zone of Nigeria agricultural soils are degrading year after year. For example, the majority of millet farmers in Kebbi and Sokoto States are experiencing this problem as a result of annual decreased production. Soils in these two States as well as in other parts of the region such as Kano States are unsustained and have less ability to support sustainable agricultural development (e.g. Harris, 2000). This problem may be attributed to unsustainable human activities on the land due to continuous cultivation or by erosion, leaching, and nutrients depletion, and or as a result of poor management. Eswarran (2001) reported that land degradation in arid, semi arid and dry areas of sub humid is resulting from various factors including climate variation and human activities. Thus, soil environment must be maintained in their natural state for the increase of agricultural production particularly for the growth of millet, cowpea and groundnut, as there are combinations of land degradation mechanisms occurring physically in the zone.

Researcher indicates that the mechanisms that initiate land degradation include physical, chemical and biological process (Lal, 1994). However, important among physical process are a decline in soil structure leading to crusting, compaction, erosion, desertification, anaerobism, environmental pollution, and unsustainable use of natural resources (Eswarran *et al.,* 2001). Significant chemical processes include acidification, salinization, decrease in cation retention capacity, and fertility depletion and Biological processes include reduction in and biomass carbon, and decline in land biodiversity (Usman, 2007). Thus, there is a need to overcome or minimised these problems in places such as north-west zone of northern Nigeria. Possibly this can be done by adapting a sustainable soil management strategies using readily available and affordable resource, such as use of organic manure.

2.2 Key Crops growing and farming system

For many years, the key crops growing in the northwest zone of Nigeria have been millet, sorghum, maize, cowpea and groundnut. Harris (2000) reported that in the sub-Sahelian part of northern Nigeria such as Gigawa, Kano, Katsina farmers focus on growing millet, sorghum, groundnut, sesame and cowpea and in the Sahelian part such as Kebbi and Sokoto they resort to the most drought-tolerant crops: millet and cowpea.

As far as farming systems are concerned, the extensive method is the predominant system farmers practice in the zone. For example State such as Kebbi and Sokoto, there are thousands of farmers growing millet and cowpea in a rotational system and sometime with sorghum (Aregheore, 2005). Still other farmers grow two or more crops before the next season e.g. millet and after the harvest, cassava in a *multiple cropping*, - the growing of two or more crops in the same land during a seasonal year (Sanchez, 1976)

Research carried out in dryland areas of northern Nigeria, the region surrounding the major state of Sokoto, Kebbi, Zamfara, Katsina, and Kano (Grove, 1962; Mortimore and Wilson, 1965) has noted three categories of farming system in the area (Mortimore, 1989): intensive system, less intensive system and extensive system. The intensive system, where permanent, annual or biannual cultivation occurs, less intensive system, where shrubs or short bush fallowing is common and the extensive system where long bush following system operates among uncultivated areas. However, soils which become subjected to these systems are usually deficient in nitrogen, phosphorus and potassium requiring high-cost improvement. Because they have become are highly susceptible to erosion, or are situated in regions affected by highly destructive exogenous geomorphologic process attributable to the climate, and therefore at increased of the risk danger of erosion (Zachar, 1982).

2.3 Environmental soil problems

The principal causes of nutrient losses from soil, which might always decreases crop yield in northwest zone may be attributed to soil erosion and land degradation by leaching, desertification, deterioration of soil structure and loss of organic matter. However, research carried out in the sub-Saharan Africa (Stoorvogel and Smaling, 1990), highlighted concerns over soil nutrient depletion in the region. Mango (1996) indicates that the principal causes of nutrient losses are crop harvest, leaching, run-off and soil erosion. This view, however, differs from that of Van Dar Pol, (1992) who noted that soil fertility is declining due to inadequate use of management input in agricultural soil among sahelian farmers. When these causal effects combine, however, as an integrated factor, occurring in a particular region will no doubt take a greater part in causing a serious environmental soil problem that might leads to decreases in crop yield.

Factors increasing these problems are drought and poor land use management and the expected more extreme climatic condition due to climatic change (ISSDS, 2001). The study of the kinetic energy of rains at Samaru Zaria in the northern Nigeria has further indicated that raindrops accomplish much erosional and transportation work (Kowal and Kassan, 1977). Similar previous studies on kinetic energy of raindrops have shown the existing of this occurrence (e.g. Hudson, 1971; Jansson, 1982). Other causes include drought, population pressure, failure to implement appropriate technologies, and poverty (Virmani *et al., 1994*). These factors could serve as contributing to ongoing soil nutrients depletion in the northwest of northern Nigeria. People are destroying forest trees and shrubs for agricultural activities or as a source of fire wood. As matter of fact, when forests are destroyed, no human power can control the destructive force of the torrential rains that sweeps the unprotected soil from (damaging) (Whyte and Jacks, 1986). Emphasising on this, two most associated problems, physically seen in the context of field soil environment have been further discussed in the following paragraphs – soil erosion and desertification.

2.3.1   Soil Erosion and causal Agents
Soil erosion can be defined as detachment and deposition of soil particles by wind, water gravity and or by other forces from one place to another (Datta, 1986). The rate of soil loss (by erosion) is normally expressed in units of mass or volume per unit of time (Morgan, 1986).

In the northwest zone of northern Nigeria, the amount of soil erosion is largely occurring not only by soil itself, but also by the management practices, usually receives using tractors or other mechanical applications almost every year. Similar cases were also reported by Datta (1986) from India. Other contributing factors to this problem in the zone may also include overgrazing, overexploitation of forest land and natural vegetation. According to Whyte and Jacks, (1986) when natural grassland, on even the gentlest slope is mismanaged by injudicious cultivation, say, or by overgrazing, soil fertility will be reducing, erosion commences and soon the amount of run-off water begins to increase. Hudson (1971) shows that erodibility is influenced by the management than by any other factors, and management includes both the broad issues of land management and decision of crop management.

Two agents of erosion might be attributed to northwest zone of Nigeria: water and wind erosion. These are also the main causal agents of erosion in countries such as India (Datta, 1986); North and Southern part of America (Hudson, 1982); Southern Ethiopia (Elias, 1998); and Kenya (Mango, 1999). Hudson (1982) mentioned: for water erosion country affected includes America, nearly all of Africa except the dry deserts and equatorial forest, Asia and Australia excluding the dry centre. Areas for wind erosion covered north America (the Great plains, famous as the Dust Bowl), the Sahara and Kalahari Desert in Africa, Central Asia (Particularly the steppers of Russia), and Central Australia (Hudson, 1982).

2.3.2 Desertification

Another environmental problem causing massive hazard to soil and decrease in crop yield in northwest zone of Nigeria is desertification. The most accepted definition of desertification as one time declared during the 2001 international conference on land degradation and desertification in Thailand was that of "Dregne" who defined it as "The impoverishment of terrestrial ecosystems that can be measured by reduced productivity of desirable plants, undesirable alterations in the biomass and the diversity of the micro and macro flora and fauna, accelerated soil deterioration, and increased hazards for human occupancy" (ICLDD, 2001).

Physically, the problem of desertification is increasingly occurring in the northwest zone of northern Nigeria as a result of forest destruction where timber still remain as the major source of cooking materials. This problem has covered a wide area not only in northwest zone of northern Nigeria but in many part of the world. During the international conference on land degradation and desertification (ICLDD, 2001) in Thailand it was reported that "the forest consumption is about 3 million ha per annum". This might be as a result of several factors but in sub-Saharan Africa the concern is population growth without better living standard, poor agricultural annual yield performance and environmental degradation (Cleaver and Schreiber, 1994). Other factors have been mentioned by Grove (1997) in his book named *"Desertification in the African Environment"*

Conclusion remarks

It is believe that a better understanding of agricultural soil environment is a key to better soil quality and high crop yield. Understanding the soil problems would certainly leads to solutions on how to tackle them. However, this could not only serve as a part of soil management but also as a part of economic development. For instance, when mother is healthy and fertile, there is more expectation from her to provide healthy and quality produce but when there is deficiency in any of the elements (either macros or micros elements), her produce may also affected, hence increasing poverty and hunger. High quality and fertile soil always provide better condition for the growth and development of plants (right from seed germination up to the maturity stage). Example, deficiency in N, P, and K in soil might lead to several problems to growing plant in a particular soil environment. These problems may include poor germination, stunting, and defoliations of plants, late maturity and also susceptible to both pest and disease attack leading to poor yield of the crops. Therefore, understanding soil environment is a key to high yield of crop production in the northwest one of northern Nigeria. The following should be kept in mind as undergoing agricultural production in the region:

- applying organic matter regularly,
- growing cover crops in rotation with millet/sorghum
- intercropping and multiple cropping systems
- shifting cultivation if possible
- crop rotation system
- planting trees (e.g. Eucalyptus spp.) around the farm
- minimum tillage system
- Planting legumes in the early season (first rainfall) and after it fully germinated, it will then follows with land preparation by turning it up and down. This is important in soils deficient in nitrogen or low organic matter
- Also good government system is important
  Many of these practices has been reviewed in Mango, (1996); Millar, (1963); Usman, 2007 etc.

Reference

Adegbola, S. A. (1979). An Agriculture Atlas of Nigeria, oxford, University press, Oxford. Cited in Aregheore, E. M. (2005). Country pasture/forage Resource Profiles: Nigeria. University of South pacific, school of Agriculture, Apia, SAMOA.

Aregheore, E. M. (2005). Country pasture/forage Resource profiles: Nigeria. University of South pacific, school of Agriculture, Apia, SAMOA. http://www.fao.org/ag/AGPC/doc/counprof/nigeria.htm

Cleaver K. M. and Schreiber, A. G. (1994) Reversing the spiral, World Bank, Washington, D. C. (Cited in Federal Ministry of Environment Report, 2002)

Datta, S. K. (1986) Soil conservation and Land management. International book Distribution, Society Ltd. Dehra Dun India pp5&18

D'Hoore, J.L. (1964) *Soil map of Africa (Scale: 1 to 5, 000,000)*. Joint project no. 11, Commission for Technical Co-operation in Africa, Lagos. Essiet, E.U. (n.d.) Unpublished data. Cited in Harris, F.M.A. 2000. Journal of Experimental Agriculture.

Elias, E. (1998) *"Is soil fertility decline?"* Perspectives on environmental change in Southern Ethiopia. Managing Africa's soils No. 2. Dryland programme, IIED, Endsleigh street London. p3.

Eswaran, H., Lal R. and Reich, P. F. 2001. An overview: *"Land degradation"* International Conference on Land Degradation and Desertification, Khon Kaen, Thailand. Oxford Press, New Delhi, India.

Fitzpatrick, E.A. (1980) Soils: Their Formation, Classification and Distribution. Printed in Great Britain at the Printman press, Longman Group Ltd. London. 138-140.

FMEN, (2001) Federal Ministry of Environment of Nigeria: *National action program To combat desertification.* Federal Ministry of Environment, Abuja, Nigeria.

Gadzaama, N. M. (1997) Desertification in the Arid Zone of Nigeria. Monograph Series No.1, Centre for Arid Zone Studies, University of Maiduguri, Nigeria. 33pp.

Grove, A. T. (1997): *"Desertification in the African Environment"* Centre for African Studies, London.

Grove A.T. (1962) "Population densities and agriculture in northern Nigeria", in Barbour, K.M. and Prthero, R.M. *Essay on African population,* 115-136. Routledge and Kegan Paul London.

Harris, F.M.A. (2000) "Manure management by smallholder farmer in the Kano Clossettled zone", submitted to *Journal of Experimental Agriculture.* Nigeria.

Hudson, N.W. (1981) Soil Conservation. Bastford Academic and Educational Ltd. 4 Fitzhard Street, London. 27-30, 86.

Hudson, N. W. (1971) Soil Conservation. Bastford Ltd, Lond. Instituto National de Technology Agropercuaria, Degradation de loss Suclos.

ICLDD, (2001). Second ICLDD: International Conference on Land Degradation and Desertification. Kohon Kaen, Thailand. Oxford press, New Delhi, India.

Iloeje, N. P. (2001). A new geography of Nigeria, New Revised Edition, Longman Nigeria PLC. 200p.

ITTA (1997). Farmers' perceptions of soil degradation. Annual report 1997. ITTA, Ibadan, Nigeria.

ISSDS, (2001). International conference on sustainable Soil Management for Environmental protection, Soil physical aspects, University of Florence, Italy. Istituto Sperimental per Lo Studio e la Difessa Del Suolo. http//www.issds.it/comference

Janson, M. B. (1982). Land erosion by Water in different climates. Department of Physical Geography, Uppsala University. UNGI Report No. 57. Borgstronns Trykeri AB, Motala, Sweden.

Jones, M. and Wild, A. (1975) "Soils of the West Africa savannah" Technical Communication, No.55. Commonwealth Bureau of Soils. Commonwealth Agriculture Bureau, Harpenden. In Harris, F. M. A. 2000.

Kiowa, J. M. and Kassan, A. H., (1977). Energy Lord and Instantaneous intensity of Rainstorms at Samaru, Northern Nigeria. In Soil Conservation and Management in the humid tropics (Ed.) Greenland, D. J. and Lal. R., Wiley, Chichester.

Lal, R. 1994. Tillage effects on soil degradation, soil resistance, soil quality, and Sustainability. *Soil tillage research,* 27, 1-8.

Mango, N. A. R. (1996). Integrated Soil fertility management in Siaya District, Kenya. Managing Africa's Soils No. 7. IIED, Eilleen Higgin, Russell Press. Dryland Programme, IIED, 3 Endsleigh street London. p1.

Millar, C. A. (1963). Soil fertility. John Wiley & Sons Inc. United state of America. Pvii.

Morgan, R. P. C. (1986). Soil Erosion and Conservation edited by Davidson. D. A. University of Strathclyde, Longman Scientific and technical Ltd. Hong Kong.

Mortimore, M. J. (1989). Adapting to Drought, Farmers, Famine and Desertification in West Africa, Cambridge University Press. (Cited in Federal Ministry of Environment Report, 2002)

Mortimore, M. and Wilson, J. (1965). "Land and people in the Kano close-settled Zone", *Occasional paper 1.* Department of Geography, Ahamadu Bello University, Zaria (ABU), Nigeria.

Russell, J. S. and Greacen E. L. (1977). Soil factors in crop production in a Semi- arid Environment. University of Queensland press in association with Australian Society of soil science incorporated, University of Queensland press. p.3.

Sanchez, P. A. (1976). Properties and Management of Soils in the Tropics. Department Of soil Science, North Carolina state University, John Wiley and Sons. 478pp.

Stoorvogel, J.J., Smaling, E.M.A and Jansen, B.J. (1993). Calculating soil nutrient balances in Africa at different scales. I. Supra-national scale. *Fertilizer Research, 35, 227-235*

Stoorvogel, J.J. and Smaling, E.M.A. (1990). *Assessment of soil nutrient depletion in Sub-Saharan Africa, 1983-2000.* Report 28, Winand Centre for integrated Land, Soil and Water Research (SC-DLO), Wageningen, Netherlands.

USDA, (1975). Soil Taxonomy, Agricultural Hand book No. 436, pp754.

Usman, S. (2003). Determination of Infiltration rate in Sokoto Rima flood basin. The Undergraduate research project, supervise under the guide of Head of soil Science department, Faculty of Agriculture, university press. Usman Danfodiyo University Sokoto, Nigeria.

Usman, S. (2007) Sustainable soil management in the dryland of northern Nigeria. GRIN Publishing, Germany.

Usman, S. (2013) Understanding soil: environment and property under agricultural condition. PublishAmerica United State of America.

Van Der Pol, F. (1992). "Soil mining. An unseen contributor to farm income in Southern Mali", *Royal Tropical Institute Bulletin 325.* Amsterdam.

Virmani, S.M., Katyal, J.C., Eswaran, H. and Arrol, I., Eds, (1994). Stressed Agroecosystem and sustainable Agriculture. New Delhi. Oxford and IBH.

Whyte, R. O. and Jacks, G. V. (1982). The rape of the Earth: A world Survey of soil Erosion. Faber and Faber Ltd. 24 Russell Square

London, Seventh Impression march MCMLVI, Great Bretain, Maclehose and Company Ltd. The University Press Glasshow.

Wild, A. (1993). Soils and Environment: An Introduction. Cambridge university press Trumpington Street, Cambridge New York USA. p88.

YOST, R., Uehara, G., Itoga, S. and Puta, G W. A. (1987). Application of Expert Systems to the transfer of soil management technology. University of Hawaii And centre for Research – In TropSoils: Technical Report 1985-1986. 177pp.

Zachar, D. (1982). Soil erosion. Development in soil science 10, forest Research Institute, zoolen, Czechoslovakia. Else vier scientific publication company. Amsterdam, Oxford, New York. p9.